FOREWORD

In a world where dreams ignite the flames of innovation and adrenaline veins of speed seekers, there exists a realm of automotive marvels known as supercars. These magnificent machines transcend ordinary limits, leaving behind a trail of wonder and awe wherever they roar.

Welcome, dear readers, to a coloring adventure that will transport you to the heart of this thrilling universe. As you open this book, you're about to embark on an exhilarating journey filled with art, knowledge, and boundless creativity.

At the heart of every supercar lies a story—a tale of visionary engineers, passionate designers, and the relentless pursuit of perfection. These pages hold a gallery of speed and style, each car a masterpiece in its own right, carefully curated to take you on a whirlwind of astounding human achievement.

Feel the pulse of raw power as you color the sleek lines of the Bugatti Veyron, a legend that once reigned as the fastest production car in the world. Sense the elegance and grace of the Lamborghini Aventador, an artist's canvas come to life in the form of automotive excellence.

But this coloring book is more than just a tribute to these mechanical marvels. With each turn of the page, immerse yourself in unique scenic settings that breathe life into these automotive stars. Picture the roaring engines echoing through sun-kissed canyons, or imagine the wind rushing through your hair as you cruise along coastal roads at sunset. Every backdrop has been thoughtfully paired to capture the essence of the supercar and set the stage for your artistic exploration.

As you bring these cars to life with a spectrum of colors, you'll also find performance statistics and intriguing facts. Delve into the depths of technological prowess, and discover the stories that make each supercar an icon in its own right.

Thank you for choosing Creative Journey Press as your companion on this artistic journey. I hope you will find inspiration, joy and intrigue as you delve into the pages ahead. We thrive off of feedback and would appreciate if you could take a moment to leave an honest review. This not only motivates us to create more captivating coloring journeys, but also helps other fellow adventurers discover the magic that awaits within these pages.

Wishing you countless colorful adventures ahead,

M.T. Eddie
Founder, Creative Journey Press

PAGANI ZONDA

The Pagani Zonda was named after a wind that blows over the Andes mountains in South America. In fact, the design of the car was inspired by the natural curves and shapes of the wind-swept mountains. The founder, Horacio Pagani, is known for his obsession with carbon fiber, and the Zonda was one of the first cars to extensively use the lightweight material in its construction. As a result, the Pagani Zonda is not only a stunning work of automotive design but also a technological masterpiece that pushes the limits of what's possible in a road car.

- Top Speed: 233 mph (375 km/h)
- 0-60 mph (0-97 km/h): 2.7 sec
- Engine: 6.0-liter V12
- Horsepower: 800 hp
- Weight: 2,359 lbs (1,070 kg)
- Braking: Carbon ceramic brakes
- Production: 1999-2017 (140 units)
- Cost in 2023: 280K to 17.5M

PORSCHE 918 SPIDER

The Porsche 918 Spyder is not only a powerful hybrid hypercar, but it also holds the distinction of being the first street-legal production car to break the 7-minute barrier at the Nürburgring Nordschleife. With a lap time of 6 minutes and 57 seconds, the 918 Spyder showcased its exceptional performance and track prowess. This achievement highlights the car's advanced hybrid technology, remarkable speed, and precision handling, solidifying its place as a true automotive masterpiece.

- Top Speed: 214 mph (345 km/h)
- 0-60 mph (0-97 km/h): 2.5 sec
- Engine: 4.6-liter V8 engine combined with 2 electric motors
- Horsepower: 887 hp
- Weight: 3,602 lbs (1,635 kg)
- Production: 2013-2015(918 units)
- Cost in 2023: 845K to 2M

APOLLO IE

The Apollo Intensa Emozione (IE) supercar is known for its intense focus on delivering a pure and exhilarating driving experience. The car showcases dramatic aerodynamic elements, including a large rear wing and aggressive body lines, which not only enhance its performance but also create a visually captivating presence. The Apollo IE's design is aimed at maximizing downforce and providing exceptional handling dynamics, making it a standout among supercars in terms of both aesthetics and performance.

- Top Speed: 208 mph (335 km/h)
- 0-60 mph (0-97 km/h): 2.7 sec
- Engine: 6.3-liter naturally aspirated V12
- Horsepower: 769 hp
- Weight: 2,866 lbs (1,300 kg)
- Production: 2018+ (10 units)
- Cost in 2023: 2.67M

ASTON MARTIN ONE-77

With only 77 units in production, the Aston Martin One-77 is highly sought after by collectors and enthusiasts. The One-77 is known for its meticulous craftsmanship and attention to detail. Each car was hand-built and took approximately 2,700 hours to produce, showcasing Aston Martin's commitment to luxury and precision. Its design incorporates advanced aerodynamics and a powerful V12 engine, making it a remarkable combination of beauty and performance.

- Top Speed: 220 mph (354 km/h)
- 0-60 mph (0-97 km/h): 3.5 sec
- Engine: 7.3-liter naturally aspirated V12
- Horsepower: 750 hp
- Weight: 3,594 lbs (1,630 kg)
- Production: 2009-2012 (77 units)
- Cost in 2023: 1.6M

SHELBY COBRA

The Shelby Cobra was originally designed as a performance car to challenge the dominance of European sports cars in the 1960s. By combining a lightweight British AC chassis with a powerful American V8 engine, Carroll Shelby created a formidable and distinctive sports car. The Shelby Cobra's success on both the road and the racetrack solidified its place in automotive history, as it not only achieved remarkable speed and power but also revolutionized the perception of American sports cars on a global scale.

- Top Speed: 150 mph (241 km/h)
- 0-60 mph (0-97 km/h): 4-6 sec
- Engine: V8
- Horsepower: 300-600hp
- Weight: 2,000 to 2,500 pounds (900 to 1,100 kg)
- Production: 1962-1967 (998 units)
- Cost in 2023: 1M

DODGE VIPER VX

The Dodge Viper, known for its raw power and iconic design, has an interesting historical tie to the famous automotive designer Carroll Shelby. The concept for the Viper was actually initiated by Shelby and a group of Chrysler executives during a discussion at a car show in 1988. Shelby, known for his involvement with legendary vehicles like the Shelby Cobra, played a significant role in the early development and design of the Viper. This connection to Shelby adds to the Viper's rich heritage and the legacy of American muscle cars.

- Top Speed: 206 mph (331 km/h).
- 0-60 mph (0-100 km/h) 3.5 sec.
- Engine: 8.4-liter V10 engine.
- Horsepower: 645 horsepower.
- Weight: 3,400 lbs (1,542 kg).
- Production: 2013-2017 (3735 units)
- Cost in 2023: 80K - 120K

BENTLEY CONTINENTAL GT3-R

The Bentley Continental GT3-R is the most dynamic and powerful road car ever produced by Bentley at the time of its release. It showcases a perfect combination of luxury and high-performance, with extensive weight reduction, improved aerodynamics, and enhanced power output. The GT3-R's limited production run of only 300 units adds to its exclusivity, making it a highly sought-after collector's item. Its track-inspired design and exceptional performance capabilities make it a true standout in the luxury sports car segment.

- Top Speed: 170 mph (274 km/h)
- 0-60 mph (0-97 km/h): 3.4 sec
- Engine: 4.0-liter twin-turbocharged V8
- Horsepower: 572 hp
- Weight: 4,839 lbs (2,195 kg)
- Production: 2015-2016(300 units)
- Cost in 2023: 180K to 250K

BUGATTI VEYRON

The Bugatti Veyron has a unique fuel system: it has 10 radiators to cool down its massive engine, but the car's fuel tank takes just 12 seconds to empty completely when driven at its top speed. To prevent fuel starvation, the Veyron utilizes a dual fuel pump system that can pump fuel at an astonishing rate of 26.4 gallons (100 liters) per minute, ensuring a constant and uninterrupted fuel supply even under extreme speed and acceleration.

- Top Speed: 267 mph (431 km/h).
- 0-60 mph (0-97 km/h): 2.5 sec
- Engine: Quad-turbocharged 8.0-liter W16 engine.
- Horsepower: 1,001 horsepower.
- Weight: 4,162 lbs (1,888 kg).
- Production: 2005-2015(450 units)
- Cost in 2023: 1.9M

PAGANI HUAYRA

The Pagani Huayra features an innovative active aerodynamics system called "AeroFlaps." These four independently controlled flaps on the front and rear of the car dynamically adjust to optimize downforce and reduce drag. During cornering, the AeroFlaps on the inside of the turn rise to increase downforce on that side, enhancing stability. This system ensures that the Huayra maintains exceptional handling and stability at high speeds, delivering an exhilarating driving experience while maintaining a stunning aesthetic appeal.

- Top Speed: 230 mph (370 km/h.)
- 0-60 mph (0-100 km/h) 2.8 sec
- Engine: Twin-turbocharged 6.0-liter V12 engine
- Horsepower: 730 horsepower.
- Weight: 2,976 lbs (1,350 kg).
- Production: 2011-2018 (100 units)
- Cost in 2023: 2M - 4M

FERRARI F40

The Ferrari F40, introduced in 1987, was the last car personally approved by Enzo Ferrari. It was built to commemorate Ferrari's 40th anniversary and became an icon of automotive design. With a top speed of 201 mph (324 km/h), it was the first street-legal production car to exceed 200 mph. The F40 was also one of the first production cars to feature a tubular steel frame with composite panels, contributing to its lightweight construction and exceptional performance.

- Top Speed: 201 mph (324 km/h).
- 0-60 mph (0-100 km/h): 3.8 sec
- Engine: Twin-turbocharged 2.9-liter V8 engine.
- Horsepower: 471 horsepower.
- Weight: 2,425 lbs (1,100 kg).
- Production: 1987-1992 (1300 units)
- Cost in 2023: 1.5M - 2.5M

ZENVO ST1

The Zenvo ST1, a Danish hypercar, is known for its extreme performance and unique features. It features an innovative Centripetal Wing that adjusts its angle based on the car's speed and lateral forces, optimizing downforce and stability during high-speed cornering. Additionally, the ST1's 6.8-liter V8 engine is paired with both a supercharger and a turbocharger, resulting in a staggering output of 1,104 horsepower. These elements contribute to the ST1's reputation as an exceptional and attention-grabbing hypercar.

- Top Speed: 233 mph (375 km/h).
- 0-60 mph (0-100 km/h) 2.9 sec.
- Engine: 6.8-liter V8
- Horsepower: 1,104 horsepower.
- Weight: 3,027 lbs (1,375 kg).
- Production: 2009-2016 (15 units)
- Cost in 2023: 1.2M - 1.8M

FERRARI F12 TDF

The Ferrari F12 TDF is a limited-edition variant of the F12berlinetta, built as a tribute to the Tour de France automobile race. It features many enhancements over the standard F12, including increased power, reduced weight, and improved aerodynamics. The F12 TDF also has an active rear-wheel steering system, which enhances stability and agility. At low speeds, the rear wheels turn in the opposite direction to the front wheels, while at higher speeds, they turn in the same direction, resulting in improved cornering performance and responsiveness

- Top Speed: 211 mph (340 km/h).
- 0-60 mph (0-100 km/h) 2.9 sec.
- Engine: 6.3-liter V12 engine.
- Horsepower: 769 horsepower.
- Weight: 3,120 lbs (1,415 kg).
- Production: 2015-2017 (799 units)
- Cost in 2023: 1.5M to 2.5M

FERRARI 599 GTO

The Ferrari 599 GTO holds the distinction of being the fastest road-legal Ferrari at the time of its release. It achieved this feat by lapping the Fiorano test track in just 1 minute and 24 seconds, faster than any other road-going Ferrari. Its aerodynamic enhancements, weight reduction, and powerful V12 engine make it a formidable performance machine, delivering an exhilarating driving experience. With only 599 units produced, it remains a highly coveted and exclusive model among Ferrari enthusiasts and collectors.

- Top Speed: 208 mph (335 km/h).
- 0-60 mph (0-100 km/h) 3.3 sec.
- Engine: 6.0-liter V12 engine.
- Horsepower: 661 horsepower.
- Weight: 3,615 lbs (1,640 kg).
- Production: 2010-2012(599 units)
- Cost in 2023: 700K-900K

FERRARI 488 PISTA

The Ferrari 488 Pista is not only powerful but also lightweight. It incorporates an array of weight-saving measures, including a carbon fiber body and a stripped-down interior. The 488 Pista's advanced aerodynamics, derived from Ferrari's racing experience, generate impressive downforce for enhanced stability and cornering capabilities. This combination of power, lightweight construction, and aerodynamic efficiency makes it an exceptional track-focused supercar.

- Top Speed: 211 mph (340 km/h).
- 0-60 mph (0-100 km/h) 2.8 sec.
- Engine: Twin-turbocharged 3.9-liter V8 engine.
- Horsepower: 710 horsepower.
- Weight: 2,822 lbs (1,280 kg).
- Production: 2018-2020 (3500 units)
- Cost in 2023: 350K to 450K

LAMBORGHINI VENENO

The Lamborghini Veneno, a limited-edition hypercar, is known for its extreme design and exclusivity. With only five units produced, it is exceptionally rare. The Veneno draws inspiration from stealth fighter jets, featuring sharp lines, aggressive aerodynamics, and a powerful 6.5-liter V12 engine producing 740 horsepower. The Veneno's unique and dramatic styling, combined with its outstanding performance, makes it an automotive masterpiece and a symbol of Lamborghini's boundary-pushing engineering.

- Top Speed: 221 mph (355 km/h).
- 0-60 mph (0-100 km/h)2.8 sec.
- Engine: 6.5-liter V12 engine.
- Horsepower: 740 horsepower.
- Weight: 3,285 lbs (1,490 kg).
- Production: 2013-2014 (5 units)
- Cost in 2023: 11M

SALEEN S7

The Saleen S7, an American mid-engine supercar, had an interesting feature called "Butterfly Doors." These unique upward-opening doors not only added to the car's dramatic styling but also had a practical purpose. Unlike conventional doors, the butterfly doors allowed easier access and exit from the vehicle in tight parking spaces or crowded areas, making it a distinctive and functional element of the S7's design.

- Top Speed: 220 mph (354 km/h)
- 0-60 mph (0-100 km/h)2.8-3.3 sec.
- Engine: 7.0-liter V8 engine.
- Horsepower: 550-750+
- Weight: 2,750-2,950 lbs (1,250-1,340 kg)
- Production: 2000-2009
- Cost in 2023: 500K to 1M

MERCEDES-BENZ SLS AMG BLACK SERIES

The Mercedes-Benz SLS AMG Black Series, a high-performance variant of the SLS AMG, holds the title as the most powerful production car ever developed by AMG at the time of its release. The Black Series features extensive weight reduction, aerodynamic enhancements, and a track-focused suspension, resulting in exceptional performance and handling. With limited production numbers, it remains a highly sought-after and collectible sports car.

- Top Speed: Electronically limited to 196 mph (315 km/h).
- 0-60 mph (0-100 km/h) 3.5 sec.
- Engine: 6.2-liter V8 engine.
- Horsepower: 622 horsepower.
- Weight: 3,417 lbs (1,550 kg).
- Production: 2013-2014 (50 units)
- Cost in 2023: 400K to 600K

PORSCHE CARRERA GT

The Porsche Carrera GT features a unique and high-revving V10 engine derived from Porsche's endurance racing program. It was one of the first production cars to utilize a monocoque chassis made entirely of carbon fiber reinforced polymer (CFRP), resulting in a lightweight yet incredibly strong structure. The Carrera GT's iconic design, superb handling, and exhilarating driving experience make it a highly sought-after collector's car among automotive enthusiasts.

- Top Speed: 205 mph (330 km/h).
- 0-60 mph (0-100 km/h) 3.9 sec..
- Engine: 5.7-liter V10 engine.
- Horsepower: 612 horsepower.
- Weight: 3,043 lbs (1,380 kg).
- Production: 2004-2007 (1270 units)
- Cost in 2023: 1.3M

MERCEDEZ-BENZ CLK-GTR

The Mercedes-Benz CLK-GTR was originally built for racing in the FIA GT Championship. In order to meet the homologation requirements, Mercedes-Benz had to produce a limited number of road-going versions of the CLK-GTR. Only 25 road cars were built, making it an extremely exclusive and sought-after model. The CLK-GTR featured a powerful V12 engine, aerodynamic design, and advanced racing technology, solidifying its place as one of the most legendary and collectible Mercedes-Benz cars ever produced.

- Top Speed: 214 mph (345 km/h).
- 0-60 mph (0-100 km/h) 3.8 sec.
- Engine: 6.9-liter V12 engine.
- Horsepower: 604 horsepower.
- Weight: 2,866 lbs (1,300 kg).
- Production: 1997-1998(25 units)
- Cost in 2023: 8.5M

KOENIGSEGG THE ONE

The Koenigsegg One:1, known as "The One," was not just a hypercar but a technical marvel. It was the first production car to achieve a power-to-weight ratio of 1:1, with 1,360 horsepower and a weight of 1,360 kg. This extraordinary balance enabled mind-boggling acceleration and performance. With only seven units produced, each meticulously crafted and personalized, "The One" represents the epitome of Koenigsegg's engineering excellence and exclusivity.

- Top Speed: 273 mph (439 km/h).
- 0-60 mph (0-100 km/h) 2.8 sec.
- Engine: 5.0-liter twin-turbocharged V8 engine.
- Horsepower: 1,360 horsepower.
- Weight: 1,360 kg (2,998 lbs).
- Production: 2014-2015(7 units)
- Cost in 2023: 7.2M

MCLAREN 720S SPIDER

The McLaren 720S Spider, a convertible supercar, features an innovative retractable hardtop that can be operated even while driving at speeds of up to 31 mph (50 km/h). This allows for the exhilarating experience of open-top driving at any moment. With its lightweight carbon fiber construction, powerful twin-turbocharged V8 engine, and advanced aerodynamics, the 720S Spider offers impressive performance and dynamic handling.

- Top Speed: 212 mph (341 km/h).
- 0-60 mph (0-100 km/h) 2.8 sec.
- Engine: 4.0-liter twin-turbocharged V8 engine.
- Horsepower: 710 horsepower.
- Weight: 3,350 lbs (1,520 kg).
- Production: 2017-2023 (765 units)
- Cost in 2023: 335K

HENNESSEY VENOM GT

The Hennessey Venom GT holds the Guinness World Record for the fastest production car from 2013 to 2017, reaching a top speed of 270.49 mph (435.31 km/h). It combines a lightweight chassis from the Lotus Exige with a twin-turbocharged V8 engine delivering up to 1,451 horsepower. With its aerodynamic design and incredible power-to-weight ratio, the Venom GT offers unmatched acceleration and performance. Only a limited number of units were produced, making it a rare and highly coveted supercar among enthusiasts and collectors.

- Top Speed: 270mph (435 km/h).
- 0-60 mph (0-100 km/h) 2.7 sec.
- Engine: 7.0-liter twin-turbocharged V8 engine.
- Horsepower: 1,200 to 1,451
- Weight: 2,743 lbs (1,244 kg)
- Production: 2011-2017(13 units)
- Cost in 2023: 1.2 M

AUDI R8 V10 PLUS

The Audi R8 V10 Plus is the most powerful production car Audi had ever produced at the time of its release. It features a distinctive "Virtual Cockpit" digital instrument cluster. This fully digital display replaces traditional analog gauges and provides a customizable interface, allowing the driver to prioritize relevant information. The Virtual Cockpit enhances the driving experience by offering crisp graphics, seamless integration with infotainment features, and the ability to switch between various display modes, including a full-screen map view.

- Top Speed: 205 mph (330 km/h).
- 0-60 mph (0-100 km/h) 2.9 sec.
- Engine: 5.2-liter naturally aspirated V10 engine.
- Horsepower: 610 horsepower.
- Torque: 413 lb-ft (560 Nm).
- Weight: 3,649 lbs (1,655 kg).
- Production: 2013-2018
- Cost in 2023: 200K to 300K

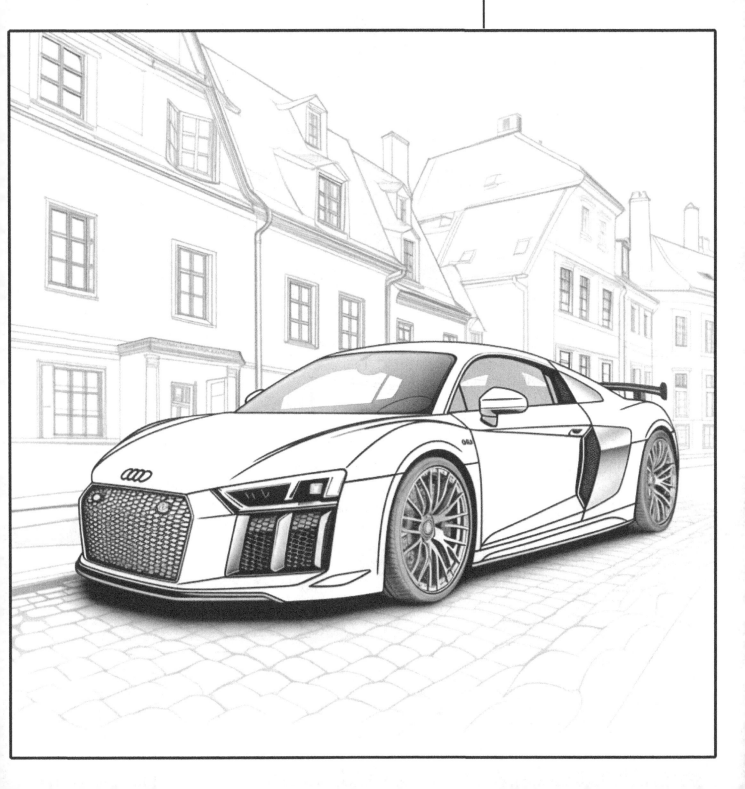

LAMBORGHINI COUNTACH

The Lamborghini Countach was not only a groundbreaking supercar but also an influential design icon. Its futuristic and aggressive wedge-shaped design, created by Marcello Gandini, set new standards in automotive styling and inspired countless generations of supercars to come. The Countach's scissor doors, sharp angles, and iconic rear wing captured the imagination of enthusiasts worldwide. Its name, "Countach," was an exclamation of astonishment in the Piedmontese dialect, further emphasizing the car's awe-inspiring presence.

- Engine: 4.0 to 5.2 liters.
- Horsepower: 375 to 455
- Top Speed: 183- 200 mph (294- 322 km/h)
- 0-60 mph (0-100km/h): 5.0-5.9 sec.
- Weight: 3,300- 3,700 lbs (1,500- 1,680 kg)
- Production: 1974-1990 (2000)
- Cost in 2023: 510K

FERRARI SCUDERIA SPIDER 16M

The Ferrari Scuderia Spider 16M was created to celebrate Ferrari's 16th Formula 1 Constructors' World Championship victory. It is a limited edition convertible version of the Ferrari F430 Scuderia, with only 499 units produced. The Scuderia Spider 16M features lightweight construction, increased engine output, and exclusive design elements. It holds the distinction of being the fastest open-top road car produced by Ferrari at the time, delivering an exhilarating driving experience combined with the renowned Scuderia performance and handling.

- Top Speed: 196 mph (315 km/h).
- 0-60 mph (0-100 km/h) 3.7 sec.
- Engine: 4.3-liter V8 engine.
- Horsepower: 503 horsepower.
- Transmission: 6-speed automated manual gearbox.
- Weight: 3,274 lbs (1,485 kg).
- Production: 2009(499 units)
- Cost in 2023: 327K

JAGUAR XJ220

The Jaguar XJ220, a legendary supercar, held the title for the world's fastest production car in the early 1990s, with a top speed of 213 mph (343 km/h). Originally envisioned with a V12 engine, it ended up featuring a twin-turbocharged V6 powerplant. Its futuristic design, innovative technology, and exceptional performance made the XJ220 a dream car for many enthusiasts. With a limited production run and striking presence, it remains an iconic symbol of Jaguar's engineering prowess and commitment to automotive excellence.

- Top Speed: 213 mph (343 km/h).
- 0-60 mph (0-100 km/h) 3.6 sec.
- Engine: 3.5-liter twin-turbocharged V6 engine.
- Horsepower: 542 horsepower.
- Weight: 3,250 lbs (1,474 kg).
- Production: 1992-1994(281 units)
- Cost in 2023: 484K

KOENIGSEGG JESKO

The Koenigsegg Jesko pushes the boundaries of automotive engineering by incorporating an innovative "Light Speed Transmission" technology, which utilizes nine forward gears and multi-clutch technology for lightning-fast gear changes. Combined with its powerful engine, aerodynamic design, and advanced features, the Jesko represents Koenigsegg's relentless pursuit of performance and showcases the cutting-edge technologies that make it a formidable presence on both road and track.

- Top Speed: 300 mph (483 km/h).
- 0-60 mph (0-100 km/h) 2.5 sec.
- Engine: 5.0-liter twin-turbocharged V8 engine.
- Horsepower: 1,603 with E85 fuel.
- Weight: 3,130 lbs (1,420 kg).
- Production: 2019(125 units)
- Cost in 2023: 3M

LAMBORGHINI AVENTADOR SVJ

The Lamborghini Aventador SVJ holds the Nürburgring lap record for production cars, completing a lap in just 6 minutes and 44.97 seconds. Its aerodynamic enhancements, including the patented ALA system, contribute to exceptional downforce and handling. The SVJ stands for "Superveloce Jota," a designation reserved for Lamborghini's most extreme and track-focused models. With its roaring V12 engine, aggressive styling, and cutting-edge technology, the Aventador SVJ represents the pinnacle of Lamborghini's performance and showcases the brand's relentless pursuit of automotive excellence.

- Top Speed: 217 mph (350 km/h).
- 0-60 mph (0-100 km/h) 2.8 sec.
- Engine: 6.5-liter naturally aspirated V12 engine.
- Horsepower: 759 horsepower.
- Weight: 3,700 lbs (1,680 kg).
- Production: 2018-2022 (800 units)
- Cost in 2023: 700K

LAMBORGHINI GALLARDO SUPERLEGGERA

The Lamborghini Gallardo Superleggera, a high-performance variant of the Gallardo, featured extensive use of lightweight materials, resulting in a weight reduction of approximately 154 lbs (70 kg) compared to the standard Gallardo. This weight-saving approach, combined with increased engine output and enhanced aerodynamics, resulted in improved performance and handling. The Superleggera's aggressive styling, distinctive rear wing, and exclusive interior trim further set it apart, making it a highly coveted and exhilarating driving experience for Lamborghini enthusiasts.

- Top Speed: 202 mph (325 km/h).
- 0-60 mph (0-100 km/h): 3.8 sec.
- Engine: 5.0-liter V10 engine.
- Horsepower: 523 horsepower.
- Weight: 2,954 lbs (1,340 kg).
- Production: 2007-2008 (172 units)
- Cost in 2023: 138K

MCLAREN P1

The McLaren P1, a remarkable hybrid hypercar, incorporated Formula 1 technology and expertise into its design. One interesting fact is that it features an Instant Power Assist System (IPAS) that provides an electric boost to the gasoline engine, resulting in a combined output of 903 horsepower. The P1's aerodynamic design, active aerodynamics, and advanced hybrid powertrain contribute to its extraordinary performance, making it a formidable competitor both on the road and on the track.

- Top Speed: 217 mph (350 km/h).
- 0-60 mph (0-100 km/h) 2.7 sec.
- Engine: 3.8-liter twin-turbocharged V8 engine combined with an electric motor.
- Horsepower: 903 horsepower
- Weight: 3,411 lbs (1,547 kg).
- Production: 2013-2015(375 units)
- Cost in 2023: 1.1M

FERRARI LEFERRARI

The Ferrari LaFerrari holds the distinction of being the first Ferrari to feature a hybrid powertrain. It combines a 6.3-liter V12 engine with an electric motor to deliver a combined output of 950 horsepower. The LaFerrari features an innovative hybrid system called HY-KERS, which utilizes regenerative braking to charge the battery and provide additional power during acceleration. This advanced technology contributes to the LaFerrari's exceptional performance, making it a true marvel of engineering and automotive excellence.

- Top Speed: 217 mph (349 km/h).
- 0-60 mph (0-100 km/h) 2.4 sec.
- Engine: 6.3-liter V12 engine combined with an electric motor.
- Horsepower: 950 horsepower
- Weight: 2,767 lbs (1,255 kg).
- Production: 2013-2016(499 units)
- Cost in 2023: 3.12M

FERRARI TESTAROSA

The Ferrari Testarossa, an iconic supercar from the 1980s, was known for its distinctive side strakes, which became one of its defining features. Interestingly, these side strakes were not merely design elements but served a functional purpose. They were designed to improve engine cooling by channeling air to the radiators located behind them, helping to dissipate heat and maintain optimal engine performance. The Testarossa's bold and innovative design, combined with its powerful V12 engine, made it a symbol of automotive style and performance during its time.

- Top Speed: 180 mph (290 km/h).
- 0-60 mph (0-100 km/h) 5.2 sec.
- Engine: 4.9-liter flat-12 engine.
- Horsepower: 390 horsepower.
- Weight: 3,800 lbs (1,724 kg).
- Design: Distinctive side strakes and pop-up headlights.
- Production: 1984-1991 (7200 units)
- Cost in 2023: 137K

MCLAREN SENNA

The McLaren Senna is named after legendary Formula 1 driver Ayrton Senna. The Senna's design was heavily influenced by aerodynamics, resulting in an aggressive and distinctive appearance. Its large rear wing, prominent air intakes, and active aerodynamic elements generate immense downforce, allowing the car to stick to the road at high speeds. The Senna represents McLaren's dedication to delivering extreme performance, precision handling, and a true driver-focused experience on both the road and the track.

- Top Speed: 208 mph (335 km/h).
- 0-60 mph (0-100 km/h) 2.8 sec.
- Engine: 4.0-liter twin-turbocharged V8 engine.
- Horsepower: 789 horsepower.
- Weight: 2,641 lbs (1,198 kg).
- Production: 2018-2019(500 units)
- Cost in 2023: 1.32M

SSC TAUTARA

The SSC Tuatara, an American hypercar, set a new top speed record for production cars in 2020. Interestingly, it achieved a staggering speed of 331.15 mph (532.93 km/h) during its record-breaking run. The Tuatara's impressive performance is attributed to its powerful twin-turbocharged V8 engine, advanced aerodynamics, and lightweight construction. Its sleek and aggressive design, coupled with its record-setting top speed, has solidified the Tuatara's position as a formidable contender in the hypercar realm.

- Top Speed: 316 mph (508.km/h)
- 0-60 mph (0-100 km/h): 2.5 sec.
- Engine: 5.9-liter twin-turbocharged V8 engine.
- Horsepower: Up to 1,750 horsepower on E85 fuel.
- Weight: 2,750 lbs (1,247 kg).
- Production: 2020+ (100 units)
- Cost in 2023: 1.9 M

PININFARINA BATTISTA

The Pininfarina Battista, an all-electric hypercar, represents Pininfarina's first foray into manufacturing its own vehicles. It is the most powerful road-legal car ever built in Italy. With a staggering 1,900 horsepower and 0-60 mph (0-100 km/h) acceleration in less than 2 seconds, it showcases the incredible capabilities of electric powertrain technology. The Battista combines luxurious design, sustainable performance, and exceptional speed, making it a true testament to the future of high-performance electric vehicles.

- Top Speed: 217 mph (350 km/h).
- 0-60 mph (0-100 km/h) 2 sec.
- Powertrain: Four electric motors, one at each wheel.
- Horsepower: 1,900 horsepower.
- Production: 2022+ (150 units)
- Cost in 2023: 2.2M

LAMBORGHINI ESSENZA SCV12

The Lamborghini Essenza SCV12, an extreme track-only hypercar, offers an exclusive ownership experience. One interesting fact is that each owner gains access to the "Lamborghini Squadra Corse Drivers Lab," a personalized training program with professional drivers and dedicated facilities. This program helps owners improve their driving skills and unleash the full potential of the Essenza SCV12 on the track. The Essenza SCV12's limited production, aerodynamic design, and powerful V12 engine make it a rare and thrilling masterpiece for Lamborghini enthusiasts seeking a track-focused experience.

- Engine: 6.5-liter naturally aspirated V12 engine.
- Horsepower: Over 820
- Top Speed: Not disclosed, as it is focused on track performance.
- Weight: 2,645 lbs (1,200 kg).
- Production: 2020 (40 units)
- Cost in 2023: 2.5M

LOTUS EVIJA

The Lotus Evija, an all-electric hypercar, is not only the first fully electric vehicle from Lotus but also the company's most powerful car to date. Its design incorporates a state-of-the-art aerodynamic system, including an innovative "Venturi" tunnel through the rear of the car that helps generate high levels of downforce. With a remarkable power output of 1,973 horsepower and a striking design, the Evija represents Lotus' commitment to pushing the boundaries of performance and electric vehicle technology.

- Top Speed: >200 mph (320 km/h)
- 0-60 mph (0-100 km/h) 3 sec.
- Powertrain: Four electric motors, one at each wheel.
- Horsepower: 1,972 horsepower.
- Production: 2019+(130 units)
- Cost in 2023: 2M

PAGANI UTOPIA

The Pagani Utopia, a new supercar, was designed with simplicity, lightness, and driving pleasure in mind. Its AMG-developed 851-hp twin-turbo V-12 engine and Carbo-Titanium core structure contribute to its lightweight nature. Weighing just 2822 pounds, it's significantly lighter than the previous Huayra Roadster. The Utopia offers a shiftable seven-speed manual or automated single-clutch transmission. With unique features like leather straps, butterfly doors, and staggered wheel sizes, the Utopia is a rare masterpiece limited to only 99 coupes produced in Italy.

- Top Speed: 220 mph (354 km/h)
- 0-60 mph (0-97 km/h): 2.6 sec
- Engine: 6.0-liter twin-turbocharged V-12 engine
- Horsepower: 851 hp
- Weight: 2,822 lbs (1,3280 kg)
- Production: 2022+ (99 units)
- Cost: 2.19M

MCLAREN SPEEDTAIL

The McLaren Speedtail, also known as the Hyper-GT, is a hybrid hypercar that combines luxury, speed, and innovation. With a top speed of 250 mph (402 km/h), it is the fastest McLaren road car ever produced. The Speedtail features a unique three-seat layout with the driver positioned centrally, surrounded by a luxurious and aerodynamically efficient interior. Its sleek and futuristic design includes active aerodynamics, retractable rearview cameras instead of side mirrors, and dihedral doors.

- Top Speed: 250 mph (403 km/h)
- 0-60 mph (0-97 km/h): 2.9 sec
- Engine: 4.0 L M840T twin-turbocharged V8 with parallel hybrid system eMotor
- Horsepower: 1036 hp
- Weight: 3153 lbs (1,300 kg)
- Production: 2020 (106 units)
- Cost: 2.25M

LAMBORGHINI MIURA

The Lamborghini Miura, an iconic sports car from the 1960s, is widely regarded as the first modern supercar. One interesting fact is that its mid-engine layout was a revolutionary design choice at the time, setting the stage for future high-performance vehicles. Its sleek and aggressive styling, along with its powerful V12 engine, made it an instant sensation. The Miura's combination of style, performance, and innovation solidified Lamborghini's position as a prominent player in the automotive world and left an indelible mark on the industry.

- Top Speed: 170 mph (274 km/h).
- 0-60 mph (0-100 km/h) 6 sec.
- Engine: 3.9-liter V12 engine.
- Horsepower: Varies 345-385
- Weight: 2,600 lbs (1,179 kg).
- Production: 1966-1973 (764 units)
- Cost in 2023: 1.7M

CHEVROLET CORVETTE C8 Z06

The Chevrolet Corvette C8 Z06, an upcoming high-performance variant of the Corvette C8, features a flat-plane crankshaft V8 engine. One interesting fact is that this engine configuration is typically found in high-performance exotic sports cars and is known for its ability to rev to higher RPMs, producing a unique and thrilling engine sound. The inclusion of a flat-plane crankshaft in the C8 Z06 showcases Chevrolet's commitment to delivering a dynamic and exhilarating driving experience in its flagship sports car.

- Top Speed: 195 mph (313 km/h)
- 0-60 mph (0-97 km/h): 2.6 sec
- Engine: 5.5-L flat-plane LT6 V-8
- Horsepower: 670 hp
- Weight: 3721 lbs (1,687 kg)
- Production: 2023 (1545 units)
- Cost in 2023: 109K

NOBLE M600

The Noble M600, a British supercar, offers a unique and engaging driving experience with a focus on driver involvement. It features a rear-wheel drive configuration and a manual transmission, making it a rarity in the era of automatic and dual-clutch gearboxes. The M600's design philosophy prioritizes lightweight construction, superb handling, and a powerful twin-turbocharged V8 engine, delivering impressive performance and a raw driving sensation that enthusiasts appreciate.

- Top Speed: 225 mph (362 km/h).
- 0-60 mph (0-100 km/h) 3.0 sec.
- Engine: 4.4-liter twin-turbocharged V8 engine.
- Horsepower: 650
- Weight: 2,755 lbs (1,250 kg).
- Production: 2010+
- Cost in 2023: 333K

MCLAREN ARTURA

The McLaren Artura introduces a new lightweight architecture called the McLaren Carbon Lightweight Architecture (MCLA), combining a lightweight carbon fiber body with a hybrid powertrain. This innovative approach enables the Artura to deliver impressive acceleration and handling while offering electric-only driving capabilities for shorter distances. The Artura represents McLaren's step towards a more sustainable future while maintaining the exhilaration and performance for which McLaren is renowned.

- Top Speed: 205 mph (330 km/h).
- 0-60 mph (0-100 km/h) 3.0 sec.
- Engine: 3.0-liter twin-turbocharged V6 engine combined with an electric motor.
- Total horsepower: 671
- Weight: 3,303 lbs (1,498 kg).
- Production: 2021+ (4662 units)
- Cost in 2023: 237K

FERRARI ROMA SPIDER

TThe Ferrari Roma is their second V-8 front-engine coupe, powered by a 612-hp twin-turbo engine and rear eight-speed dual-clutch transmission. Its powertrain roars up to 7500 rpm. Inside, the cabin features a nearly all-digital interface. Although the Roma has back seats mainly for show, the interior surprises with its spaciousness. Priced at a quarter-million dollars, this Ferrari unleashes hundreds of prancing horses, evoking butterflies in your stomach.

- Top Speed: 198 mph (330 km/h).
- 0-60 mph (0-100 km/h) 3.4 sec.
- Engine: turbocharged 3.9-L V-8
- Total horsepower: 612
- Weight: 3,713 lbs (1,684 kg).
- Production: 2023+
- Cost in 2023: 275K

FERRARI 812

The Ferrari 812 holds the distinction of having the most powerful naturally aspirated V12 engine ever produced by Ferrari. Its 6.5-liter V12 engine delivers an impressive 789 horsepower, propelling the car from 0 to 60 mph (0 to 100 km/h) in under 3 seconds. The 812 also features advanced aerodynamics and cutting-edge technology to enhance its performance and handling capabilities. With its striking design and unmatched power, the Ferrari 812 is a true testament to Ferrari's commitment to delivering exhilarating driving experiences.

- Top Speed: 211 mph (340 km/h).
- 0-60 mph (0-100 km/h) 2.9 sec.
- Engine: 6.5-liter naturally aspirated V12 engine.
- Horsepower: 789 horsepower.
- Weight: 3,594 lbs (1,630 kg).
- Production: 2017+
- Cost in 2023: 610K

FORD GT

The Ford GT pays homage to the legendary GT40 that won the 24 Hours of Le Mans four times in the 1960s. One interesting fact is that the GT features an innovative active aerodynamics system called "flying buttresses." These aerodynamic elements dynamically adjust to optimize airflow and downforce, enhancing stability and performance at high speeds. The Ford GT's combination of heritage, cutting-edge technology, and stunning design makes it a coveted and remarkable vehicle among enthusiasts and collectors.

- Top Speed: 216 mph (348 km/h).
- 0-60 mph (0-100 km/h) 3.0 sec.
- Engine: 3.5-liter twin-turbocharged EcoBoost V6 engine.
- Horsepower: 647 horsepower.
- Weight: 3,400 lbs (1,542 kg).
- Production: 2019+(4038 units)
- Cost in 2023: 500K

CZINGER 21C VMAX

The Czinger 21C Vmax features an innovative hybrid powertrain and groundbreaking 3D-printed components. It holds the distinction of being the world's first 3D-printed hypercar. This manufacturing technique allows for intricate and lightweight designs, enhancing performance and efficiency. The 21C Vmax's hybrid powertrain delivers remarkable power output and acceleration, showcasing Czinger's commitment to pushing the boundaries of automotive technology and design.

- Top Speed: 253 mph (407 km/h).
- 0-60 mph (0-100 km/h) 1.9 sec.
- Engine: hybrid twin-turbo V8 with two electric motors
- Horsepower: 1350 horsepower.
- Weight: 2,760 lbs (1,250 kg).
- Production: 2021+ (80 units)
- Cost in 2023: 2M

PANOZ ESPERANTE GTR-1

The Panoz Esperante GTR-1 holds a distinction as the first American-built car to win a major international endurance race overall. In 1997, it achieved a historic victory at the 24 Hours of Le Mans in the GT1 class, showcasing its exceptional performance and endurance capabilities. The GTR-1's unique and aggressive design, coupled with its powerful engine and aerodynamic features, contributed to its success on one of the world's most challenging race circuits, solidifying its place in motorsport history.

- Top Speed: 200 mph (322 km/h).
- 0-60 mph (0-100 km/h) 3.3 sec.
- Engine: 6.0-liter V8 engine
- Horsepower: Approximately 600 to 650 horsepower.
- Weight: Approximately 2,700 to 2,900 lbs (1,225 to 1,315 kg).
- Production: 1997-1999 (6 units)
- Cost in 2023: 890K

RIMAC NEVERA

The Rimac Nevera, an all-electric hypercar, holds the title of being one of the fastest-accelerating production cars in the world. It can sprint from 0 to 60 mph (0 to 100 km/h) in an astonishing 1.85 seconds. Its remarkable acceleration is made possible by its four electric motors and advanced torque vectoring system. The Nevera also boasts a top speed of 258 mph (412 km/h) and a range of up to 340 miles (550 km), showcasing the impressive performance and technology of electric vehicles.

- Top Speed: 258 mph (412 km/h).
- 0-60 mph (0-100 km/h) 1.85 sec.
- Powertrain: Four electric motors, one at each wheel.
- Horsepower: 1,914 horsepower.
- Torque: 1,740 lb-ft (2,360 Nm).
- Production: 221+ (150 units)
- Cost in 2023: 2M

2019 CHEVROLET CORVETTE ZRI

The 2019 Chevrolet Corvette ZR1 holds a record as the most powerful production Corvette ever made. Its supercharged 6.2-liter V8 engine produces an impressive 755 horsepower and 715 lb-ft of torque. This power enables the ZR1 to reach a top speed of over 210 mph, making it the fastest production Corvette to date. With its aggressive design, track-ready performance enhancements, and the distinctive "High Wing" option, the ZR1 represents the pinnacle of performance and engineering in the Corvette lineup.

- Top Speed: 210 mph (338 km/h).
- 0-60 mph (0-100 km/h) 2.85 sec.
- Engine: Supercharged 6.2-liter V8 engine.
- Horsepower: 755 horsepower.
- Weight: 3,560 lbs (1,615 kg).
- Production: 2019 (2953 units)
- Cost in 2023: 150K

Thank you

We hope you enjoyed this unique creative journey. Please consider taking a moment to leave us a review by scanning the QR code below.

Scan for Amazon Review

Made in the USA
Las Vegas, NV
21 December 2024

14843003R00057